W9-ACS-501

DISCARDED
from
New Hanover County Public Library

Growing Readers
Purchased with Smart Start Funds

CIRCUS SHAPES

BY STUART J. MURPHY

ILLUSTRATED BY EDWARD MILLER

NEW HANOVER COUNTY
PUBLIC LIBRARY
201 CHESTNUT STREET
WILMINGTON, NC 28401

HarperCollins*Publishers*

LEVEL 1

To Samantha—the first of the next generation
—S.J.M.

To Kristen Lynn Cox
—E.M.

The illustrations for this book were created on the computer.

Bugs incorporated in the MathStart series design were painted by Jon Buller.
HarperCollins®, ®, and MathStart™ are trademarks of HarperCollins Publishers Inc.
For more information about the MathStart series, please write to
HarperCollins Children's Books, 10 East 53rd Street, New York, NY 10022,
or visit our web site at http://www.harperchildrens.com.

CIRCUS SHAPES
Text copyright © 1998 by Stuart J. Murphy
Illustrations copyright © 1998 by Edward Miller III
Printed in the U.S.A. All rights reserved.

Library of Congress Cataloging-in-Publication Data
Murphy, Stuart J. 1942–
Circus shapes / by Stuart J. Murphy ; illustrated by Edward Miller.
p. cm. (MathStart)
"Level 1, Shapes."
Summary: Circus animals and performers form basic geometric shapes as they put on a show.
ISBN 0-06-027436-0. — ISBN 0-06-027437-9 (lib. bdg.)
ISBN 0-06-446713-9 (pbk.)
[1. Shape—Fiction. 2. Circus—Fiction. 3. Stories in rhyme.] I. Miller, Edward, 1964– . II. Title.
III. Series.
PZ8.3.M935Ci 1998 96-35992
[E]—dc20 CIP
AC

Typography by Edward Miller III
2 3 4 5 6 7 8 9 10
❖

CIRCUS
SHAPES

CIRCUS

The circus is in town,
and we all get to go.

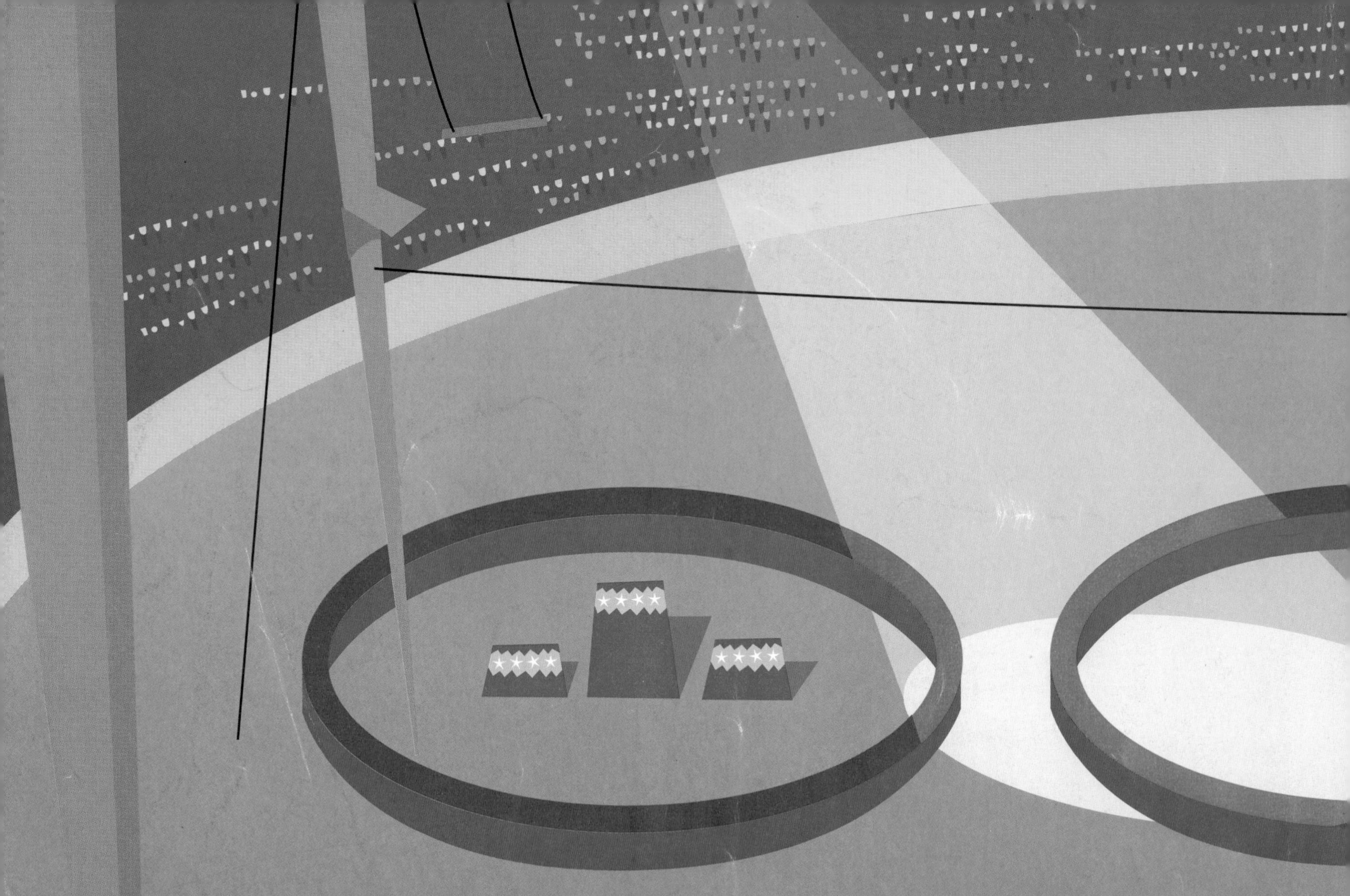

Our seats are way up high.
We're ready for the show.

The ringmaster arrives
in a funny little car.

A man on stilts walks in.
He's a favorite circus star.

Elephants make a circle
and then march 'round and 'round.

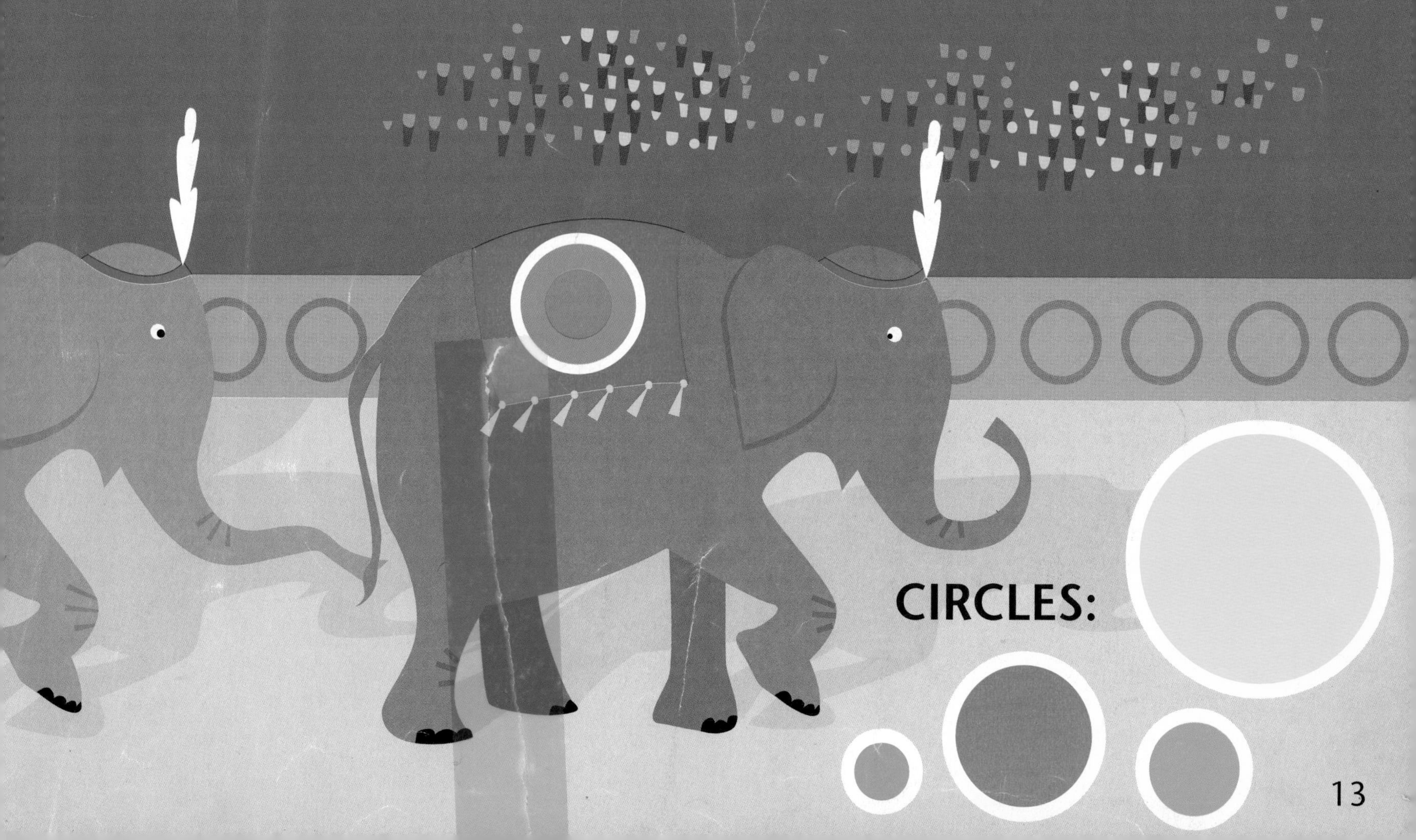

CIRCLES:

The circus band starts playing.
The tent fills up with sound.

White horses make a triangle—

three corners and three sides.

TRIANGLES:

The clowns are pulling wagons.

The dancing dogs get rides.

Some monkeys make a square.
The four sides are all the same.

The lions start their roaring.
They aren't very tame.

The bears form a rectangle—

two sides short and two sides long.

RECTANGLES:

Acrobats are twirling
to a special circus song.

Circus shapes are everywhere.

The tent is all aglow.

How many circles, triangles, squares, and rectangles can you find?

The ringmaster blows his whistle.
It's time to end the show.

FOR ADULTS AND KIDS

If you would like to have more fun with the math concepts presented in *Circus Shapes*, here are a few suggestions:

- Read the story with the child and describe what is going on in each picture. Ask questions such as "What shape are the monkeys making?" or "What shape has three sides?"

- Encourage the child to retell the story using the names of the shapes: "circle," "triangle," "square," and "rectangle."

- Look for these things around the house: faces of watches or clocks; buttons on a sweater; books, tiles, rugs, kitchen towels, and windows. Which are triangles? Circles? Squares or rectangles?

- Cut a variety of shapes from colored paper or newspapers and create pictures using the shapes. Try making a rooster, a snowman, or a dog. Create a picture of a castle, or some ice cream cones, or anything you want.

- Go on a "Shape Hunt" in your neighborhood. Create a chart like the one shown here and encourage the child to make a mark for each shape "sighting." Then add up all the marks and see how many have been found.

●	▲	■	▬
//	/////	/	///

Following are some activities that will help you extend the concepts presented in *Circus Shapes* into a child's everyday life.

Snacks: How can a sandwich be cut into squares? Into triangles? Identify the shapes of crackers and cookies. Can you bite a square cracker into a circle or a triangle? Can you bite a round cookie into a square?

Games: Cut a variety of shapes out of paper and lay them on a flat surface. Have one player cover his or her eyes while the other player removes a shape. The first player then opens his or her eyes and answers the question, "What shape is missing?"

Bedtime: Identify the shapes you see at bedtime. What is the shape of the wash-cloth, the bar of soap, or the bathroom mirror? Look at the shape of a favorite blanket or the shape of the eyes, nose, or mouth of a favorite stuffed animal. What shapes do you notice in the patterns on the sheets and pajamas?

The following stories include some of the same concepts that are presented in *Circus Shapes*:

- THE SHAPE OF THINGS by Dayle Ann Dodds
- COLOR ZOO by Lois Ehlert
- CIRCLES AND SQUARES EVERYWHERE! by Max Grover

NEW HANOVER COUNTY PUBLIC LIB.
Growing Readers
New Hanover County
Public Library
998G17 M 1076
11/04/99 166010 SELB
INFORMATION CONSERVATION, INC.